AF575996

WAYNE BINNING

Little Creek Press®
A Division of Kristin Mitchell Design, Inc.
5341 Sunny Ridge Road
Mineral Point, Wisconsin 53565

Book Design and Project Coordination:
Little Creek Press and Book Design

First Printing
January 2022

Printed in the United States of America

For more information or to order books,
www.littlecreekpress.com

Library of Congress Control Number: 2021924771

ISBN-13: 978-1-955656-14-6

Dedication

This little book is dedicated to Tanya,
my wife and best and eternal friend.
This one's for you, Babe.

Table of Contents

Brush Strokes 7
Chicken Coop Calamity 8
Christmas North of Forever 10
Degree of Kindness 11
Doctor 12
Dreamscapes 13
Farmhouse 14
First Light 16
Gophers 17
Hard Lessons 18
Heavy Heart 19
Hunter's Dream 20
Last Curtain Call 22
Husks in the Wind 24
Little Things 25
Long Shadows 26
Mom 27
October Leaves 28
Path of No Ending 29
Pillars 30
Reflections of a Christmas Soul 32
Sandbox 34
Pipe Dreams 36
Silence 37
Tadpoles and Planer Boards 38
Surf 40

Temple in the Timber 41
The Blackboard 42
The Test. 44
In a Moment 45
Keep it Simple 46
The Last Hunt 47
When Love Is Void. 48
Watch Me...Watch Me...and Keep the Yard Light On. 49
When Spirits Meet. 59
When We Are Called 60
Willow 61
Winter Wind 62
Acknowledgments 63
About the Author 66

Brush Strokes

Dedicated to Pam Willet, artist & teacher;
Inspired by one of your works

For each of us we have a different drum to beat.
We can choose a heavy trudge, or march on happy feet.
Which beat lies within an artist's choice?
Does the canvas hold a beat, or does it have no voice?

I see an ancient frigate riding on a charging sea.
I stare and wonder what message is in the hold for me.
There seems to be words of wisdom in the painter's stroke.
I have looked so many times, and with each glance
a different message spoke.

Does the message deep within the canvas hide?
Is it in the stroke of brush that subliminal truth resides?
Or is it in the stroke of brush no message is clearly told,
Merely for our eyes to rest upon, beautiful and bold?

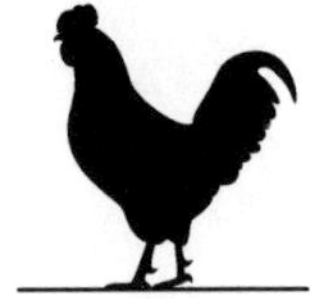

Chicken Coop Calamity

Mom kept about twenty-five hens in a coop on the farm.
On occasion, a weasel would visit causing panic and harm.
Dad got a leghorn rooster for hen protection;
Then teased and tormented him to a meanness
of perfection.

The hens were his harem and the coop was his castle.
All who entered were greeted with an attitude of no hassle.
Mom and Sis always gave the hens their water and feed;
Without a clue of his protective hostility...no indeed!

On an early, chilly morning, Sis went to feed the hens.
I was on my way to feed the pigs and clean their pens.
She left the coop door open so from the rooster she
could run.
I thought...why not close and lock the door and have
some fun!

Through the D-handle latch I put a nice little stick—
A very naughty, mean, and treacherous trick!
Locked inside, I could hear Sis scream and fight!
Feathers filled the air...what a sight.

Finally, my dad heard her scream and shout,
And came to her rescue and let her out.
She was terrified, cut, bleeding, and very teary.
Even the old rooster looked a bit weary!

I could see Dad holding that little stick in his burly,
  weathered hand.
He said he wanted an explanation upon his demand.
Suddenly, I was shaking and had every reason to be
  concerned and to worry;
In this situation, Dad was the judge and the jury.

Together, we looked at the cuts on Sis' back;
And for immediate punishment, he gave me a little
  token whack!
Then he said it was his fault too which, then,
  seemed a bit queer.
Then, I noticed upon his cheek, a tiny little tear.

It was one of many of life's lessons of regret,
But certainly nothing one could ever, ever forget.
The scars from the sharp spurs are still on Sis' back;
And though it was gentle, I still feel the sting of the
  token whack.

Christmas North of Forever

December 5, 2019

A message to my sons at Christmas

I do not know when it is that my final hour will come.
When the many miles traveled, and volumes spoken are
finally done.
A spirit piece of me has been with you always from the
very start,
Soulful thoughts of a father's love to hold within your
hearts.

A father's Christmas wish you forever made come true.
You have shown me joy, happiness, and troubles very few.
I remember when you were toddlers so very small,
And now you are grown men standing proud and tall.

There is no greater wish could come to me...
Than the wonderful sons you have grown to be.
So the bond of love between us nothing shall ever sever;
We will still celebrate Christmas and each other when
we are all North of Forever.

Degree of Kindness

December 16, 2019

"A Christmas message for Maria"

This degree of kindness is not written on any certificate or
paper form.
It is only on a tender heart where it is lovingly worn.
For a very special loving soul it was put there from
the start;
It is a kindness to hold onto for a lifetime in your
caring heart

For this kindness, no tuition fee was meant to be,
It was placed in your heart to share, priceless and free.
So when at times you are feeling low and second best,
Just share your degree of kindness and lay sadness
to its rest.

When at work you walk into a patient's room,
And all about you see and feel pain, sorrow, and gloom.
The degree of kindness in your heart shines within
the smile upon your face,
And for them in this moment their world will be a
better place.

No richer or higher degree is there to hold by anyone
on this earth,
For the gratitude and joy there is no measured worth.
So hold tight to this degree of kindness blessed within
your caring heart,
And your gentle willingness to share it with so many
others in generous parts.

Doctor

Early Winter, Sunrise
From Red, patient, Golden Retriever

A long, long time ago when we at first met,
My family told me I would need to get to know a new vet.
As I am certain you know, understand, and can tell,
My first impressions come through my wise and careful
  sense of smell.

Then it is followed with the sound of your caring voice;
From hearing it I was not afraid; this is looking like a very
  good choice.
Then I watched you move about with confidence and ease;
I could sense it was your patient you wanted most
  to please.

Then upon me was the feeling of a strong, but gentle
  touch,
I was certain these were safe hands in a time of clutch.
So, when I had to go to sleep, I really felt no primal fear;
I was comforted knowing a new and gentle friend was near.

Now I am back to have you fix my other leg like new.
I hope this surgery stuff will end with number two!
With these meetings of our eyes, this sincere story
  I have told.
Please know that your gentle, helpful kindness I will
  forever inside me hold.

Dreamscapes

Clouded memories, twisted imaginations,
Distant stories, wandering sensations.
Thoughts that confuse a tired mind;
Some are beautiful, others not so kind.

Are these premonitions of a time beyond life?
Will these thoughts then be swept free of strife?
The grasping, struggling of visions just a smear.
Will the teasing hallucinations at last become untwisted
and clear?

There was a small gift box wrapped with precision
and grace.
It was presented by a familiar, but unidentifiable face.
The smell of old and weathered cardboard from within
it seeps.
Aged and nervous fingers uncover that which inside
it keeps.

A portrait...black and white, but yellowed, it does hold.
A picture of my youthful mother...I need not be told.
Bordered by hand embroidered, yellowing lace;
However...searching, straining eyes cannot see her face.

When will these dreams...faceless dreams finally end?
Is there a secret toll to which one must first attend?
I will dream again of the portrait wrapped in lace,
And someday...once again will see her face.

Farmhouse

February 20, 2020

A temple, a shrine, a castle, my childhood resort.
A place of security, love, compassion, and peaceful
comfort.
A place where in this world I was conceived.
A place of honesty, respect, dignity...nothing was deceived.

From early morning and Dad shouting to get arising,
We all had chores to do with no complaining or
compromising.
Accountability, discipline, and commitment was our game.
Just get your job done and don't expect any fame.

Grandma's daily breakfast was our energy and our fuel.
A ritual of homemade bread, pancakes, eggs...
all before school.
Kiko, Fuzzy, and Spot would be sunning on the
east windowsill,
Waiting for leftovers for their morning fill.

We had simple games that all could play.
Along with all the work and chores, we played every day.
Dad would say, "I see something," and then we would
guess.
We had more fun than work, I must confess.

On our backs looking up at the moon and finding the
North Star;
Not a worry or a thought of our lives' future afar.
Tranquility and peacefulness let our young hearts rest,
Preparing us so every day we could do and be our
very best.

Through good and hard times, within each farmhouse wall,
We had warmth and protection so every day we could
stand up tall.
We could sleep to the scratching of mice dropping
hickory nuts.
I would not trade those days...no ifs, ands, or buts.

That farmhouse still stands with pride and grace.
And just like me, I can see the weathered years upon
its face.
It stands a monument symbol of hard and good times
of my past.
Within the walls our storied memories forever will last.

First Light

The deep and peaceful solitude of darkness is nearing the
end of another night.
This predawn moment is called by many "first light."
It marks in part an ending, and in part a new beginning.
A soul once lost, now in a day of winning.

A mind, heart, and a soul take a moment for reflection;
Scattered thoughts come together...a mindful resurrection.
Troubled times and sadness in the past that wore a frown
are washed away in first light...a healthy rebound.

Conscious thoughts once again seek answers and decisions.
First light brightens the path to new resolutions.
Thoughts lost or clouded in a muddy, stagnant pool...
are now crystal clear in first light...no longer do they fool.

In first light my gaze falls upon my companion by my side;
It is with her, my eternal friend, my complete life in her
I do confide.
It is from within her heart that shines upon me the
first light.
My paths from now through eternity will therefore
always be bright.

Gophers

January 13, 2020

Two boys, one big, the other skinny and small;
A twenty-two rifle and a dog named Teddy; we had it all.
Buster was just a pup, and our second gun was a Daisy.
Summer days were hot, cloudless, and sometimes lazy.

Gophers were the target, the game of choice.
We'd find them in the fence lines by hearing their
whistling voice.
Big brother was always in the lead...Teddy by his side.
Little brother followed close behind, but in his brother's
wisdom did confide.

Sometimes the gophers would stand up straight and tall,
Allowing for an easy shot and down they would fall.
While others streaked away and dove into their hole to hide,
Then Teddy would dig until with his jaws they would
at last collide.

These times seem like they were a long, long time ago.
And these two brothers had no way to know,
That these pesky little gophers would have the last laugh,
And that is that they are a part of our lasting epitaph.

Hard Lessons

March 3, 2020

Life is at times difficult, a struggle, filled with pain,
Testing our soul on what we want to gain.
It is tiring to worry so much of what we think we should be,
And not very important what others think of me.

It is most rewarding to know and understand who we are
than to be concerned of expectations not fitting
and bizarre.
There is a peace and comfort knowing for that which
we stand.
Once discovered and accepted, it allows a life far
more grand.

Hate is a colossal burden to carry throughout one's life.
It holds a suffering weight to bear...never-ending strife.
True forgiveness is hatred's only antidote.
It is a weightless gift within life's journey much easier
to tote.

It is within the moments of a day that our memories
mostly live.
Remember there is a lot less "take" in life, allowing
more time to "give."
Be compassionate, sincere, loving, and always kind.
Be worthy of the paths you choose...It is only your
footprints you leave behind.

Heavy Heart

December 10, 2019

Empty, hollow, darkness, a painful peal;
These are the pains a heavy heart must feel.
The soul searches for a reason of "the why."
Grasping, illusive, oblique thoughts...Still I try.

Joy and comfort, feelings of a seeming distant past.
Replaced by this oppressive burden...I will not much
longer last.
I seek an answer for the mystery guilt that I carry,
And down this path called life, I cannot afford to tarry.

Disillusioned hate and anger from a wounded, bitter soul;
I must shed this unfounded load...I will not pay the toll.
In devoted faith, kindness, and steadfast love,
I will find the help much needed from above.

I will not keep this heavy heart filled with unnecessary
sadness,
But will reload it with joyful happiness.
A healthy dose of love from friends and family from afar
will forever from this heavy heart erase this hurtful scar.

Hunter's Dream

Remembering days of seasons long, long ago,
Young hunters filled with quest would sleep forgo.
With predawn hours slow to evolve,
The patience of youth was the greatest resolve.
Upon a token opening day alarm,
A novice hunter would rise with charm.
In darkness did I trek to a chosen stand,
A place to take a trophy grand.
With little regard for this great quarry,
I did not understand his life story.
Selfish egos were my guide,
Into his world I went with determined stride.
The seasons are many that have passed,
And with them my life has changed as fast.
The grand trophy, once essential to take,
Has taught me many lessons for my own sake.
Where once from my elevated stand in a tree,
I would dream of a buck in pose for me.
He'd come from the right or maybe the left,
I'd sense him from either with eyes and ears so deft.

I was sure I was oh so wise,
To have stolen into his larder in total disguise.
But now after years of lessons learned,
It was me who he had rightly spurned.
Now it is me to whom he does speak,
With tales of life on the maple ridges and down by
 the creek.
A story he tells of a hunt gone astray,
Of how he fooled me one cloudy windy day.
On the edge of a ridge where I could both look,
Uphill or downhill by the brook.
In late afternoon when downhill are thermals,
It was most likely by the brook for life to be terminal.
A stick was broken for me to hear,
Downhill by brook so very near.
But the thermals are what I forgot,
So down by the brook he was not.
This game I've played a hundred times before,
This time the buck ran up the score.
Not fair to not play by the book,
He should have been down by the brook.
He tells me he's played this game a thousand times before,
For it is not only me he abhors.
It is a life study in his great book,
His library is on the ridge and down by the brook.
The days are many when you're not here,
Those are the times that I most fear.
The bear and the wolf on soft feet walk,
They're not as clumsy and never talk.
So next time we meet on a windy day,
Keep this in mind and may your shot go astray.

Last Curtain Call

Dedicated to Brian, my lifelong friend

The wind of summer once again waves the golden,
  nodding heads.
Gently among them the director of the harvest treads.
Skillfully the amber seeds are rolled between two
  leathery hands...
Separating seeds and chaff, an ancient test that
  forever stands.

He must now direct his reaper like a polished symphony.
All the pieces must work in unison and perfect harmony.
Like an accomplished maestro with his baton in hand,
He directs his machine to harvest still another golden stand.

He listens to the percussions, woodwinds, brass,
and the strings.
Within this massive chamber his glorious orchestra sings.
Once again he creates yet another harmonic industrial
score.
The maestro, the composer has done this countless
times before.

When upon the ending of the final movement his baton
does fall,
The maestro stands on his podium, proud and tall.
Like an artist he looks back on the picture he has painted;
He leaves his chamber, now quiet and vacated.

Was this the final show of his perfection?
Is it now the time for only memories and reflection?
Were these the last harvested seeds of fall?
Was this the maestro's final curtain call?

Husks in the Wind

The warm summer breezes are now gone.
The sun is still asleep when I awake at dawn.
The trees have dressed in colors beautiful and bright,
And the late September wind has a hint of winter's bite.

Cows in the pasture munch on summer's last grass.
Corn husks twirl in the wind and crunch like broken glass.
It's late afternoon on the old country farm.
A little girl is all dressed up in her farmgirl charm.

A blue bandanna scarf around her head...tied under
her chin,
And a missing front tooth whenever she would shyly grin.
Wearing someone's coat that was far too big;
She could barely hold it up with a body like a twig.

It was time to get the cows to the barn for milking.
Husks dancing in the wind, shed long after the silking.
Down a winding, worn cow path...not skipping too high,
Being so very careful to miss all the fragrant cow pies!

Wondering where the husks in the wind will go?
Is there a message in their twirling, dancing show?
Is there strength in something so fragile and thinned?
Maybe...just maybe, we are all husks in the wind?

Little Things

There are millions of ways to try to define true love.
Sometimes it is shown in the purity of a beautiful white
dove.
On your wedding day…a day for you most grand;
Your Olympic day…the joining of each other's hand.

Moments of happiness with friends to hold forever dear;
Memories to cling to year after year.
But it will not be these moments that define your love;
It will be a million *little things*, and help from Above.

It will be the way at each other you look.
These glances will between you write a book.
It will be thoughts of each other for no real reason;
These *little things* will be with you season to season.

It will be the magic touch of each others hand;
Only the two of you will truly understand.
It will be the fluttering feeling in your heart,
When you get back together after being apart.

It will be the inner pain for each other when you are sad;
It will be the hurt you feel when at each other you are mad.
It is all these *little things* and a million more;
That will define true love to cherish and adore.

It will be the comforting sound of each other's voice;
A sound that forever will be your number one choice.
It will be all the *little things* for which your heart sings;
It will be life's unnoticed, precious *little things*.

Long Shadows

February 27, 2020

The light of yet another day fades softly past a hazy sky.
Visions of life's hustle and bustle, darkness soon will deny.
In the midst of times of brightness and times of dark,
There falls a softer time, less obtuse, and far less stark.

This is a time when visions most often clearly seen,
begin to fade from sight with relevance not nearly as keen.
This is a time of pause, a time of mandatory reflection.
This is a time to let our soul rest...no need for perfection.

Our lives now fall within shadows cast dark and long;
These shadows are here to shelter us from harm and wrong.
The truths and lies of life within these shadows do reside.
In these long shadows we have no choice of which of these to hide.

It is now within the long shadows for a time I will stay.
I will tarry here and ponder of the toll I am certain to pay.
Upon moving from the long shadows into silence of the dark;
No longer brightness of the day...just fading singing of the lark.

Mom

No, I have not forgotten that it is nearing Mother's Day.
So, I have just a few important things to say.
I see you often in all the familiar places;
You are in my dreams that so often have no faces.

When in my garden hoeing and planting seeds,
And often too when I'm pulling weeds.
The moss roses planted just outside our door are a clue;
They were your favorite and remind me so much of you.

I want to thank you for teaching me to go potty,
and to wipe my nose when it is snotty;
How to bathe, keep clean, and comb my hair.
And to be humble and forgiving when life is not fair.

I often hear your voice and feel your touch.
It helps me because then I don't miss you as much.
Your spirit is with me and keeping me well.
Someday I'll stop by, and then we'll sit a spell.

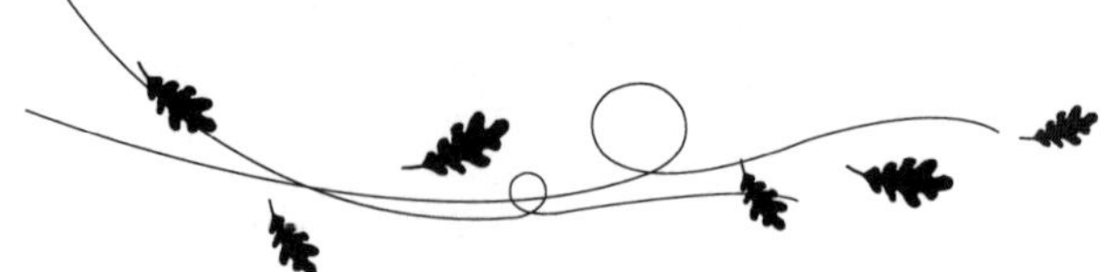

October Leaves

October 2019

The sweet winds of summer no longer kiss my cheeks,
They have succumbed to the Northers, gone again
for endless weeks.
The barren nighttime sky now holds a faceless moon,
And in the early sunset hours there is no song of a
lonely loon.

Each leaf a fading memory fluttering to the ground,
Drifting on their ending flight, whispering their final sound.
Like clouded memories, some are forever lost and gone,
While others will be rekindled with the coming of a
new spring dawn.

They all now rest in silence as the Northers bite my face,
Tenants of naked limbs have now fallen from their grace.
They stand as silhouettes against a chilling October sky,
The chorus in their crown balcony is now a moaning cry.

And still there is the painted beauty of crisp October
leaves.
Their shimmering and dancing show is not for one
who grieves.
When once again they return anew in spring,
My wish is that I am here to listen to their chorus sing.

Path of No Ending

To the love of my life, Tanya

January 16, 2020

When each other we did choose,
We set upon a journey certain not to lose.
The paths we have taken were not always smooth
  and straight;
We were determined to let our love guide our fate.

Our path took us to the woods, beautiful, dark, and deep.
It was here our hearts and souls would forever keep.
We followed on to open waters on the bay,
But it was here a crooked, bumpy path shortened our
  time of stay.

Our journey moved us onward to places of uncertainty,
Challenging both of us and love's potential frailty.
We traveled over mountains and went to places by the sea,
But it was here that the path of fate told us where to be.

So a return trek brought us back to roots and home.
It is here that upon us many blessings have been shone.
I know not of our journey's next route or if the path
  is bending,
But I am at peace because together our path has no
  unhappy ending.

Pillars

Dedicated to Dorothy & Larry on your anniversary

Two people came together; maybe it was luck or maybe it was fate.
Together they have stayed to open life's gate.
For her it was bobby socks, saddle shoes, and plaid pleated skirts.
For him it was a brown duck hunting cap, patch jeans, and faded flannel shirts.
Hardly a storied match one could say was made in heaven,
But it all got started about nineteen fifty-seven.

Their roots they had in common, both coming from the farm,
It must have been the hard country life that gave them their loving charm.
No one could have known if for each other they were ready,
But they certainly rewrote the book on a couple going steady.

Together they have stayed for over fifty years,
Building their own family and conquering life's many
troubles and fears.
Their steadiness and kind support is always there for others
to upon them lean,
Allowing others their faults of life to winnow and to glean.

Little by little, the pillars they have become, they,
themselves did build,
Not with bricks and mortar by some grand and famous
guild.
But by their hearts and character filled with kindness and
genuine love,
Faith in each other and a little help from up above.

How comforting and grand to upon your pillars all that
have gratefully leaned,
Allowing fears to be shadowed and souls to be cleaned.
These pillars will stand eternally and never sway or fall,
For our family's future they will forever stand steady
and tall.

Reflections of a Christmas Soul

It seems like only yesterday when Father Time paused at my front walk,
Said he'd like to spend some time to talk.
"How dare you stop to lecture me,
I've work not done here, can't you see?"
"Oh, work undone," he said. There'd always be."
Reluctantly, I said, "Please come on in."
With no hesitation, he said, "Where shall we begin?"
"Well, time, you know, is not of essence here,
It is not of that which you most fear.
It is of family, friends, and memories to lose,
Those are the fears that most do choose.
They are the simple choices you must make,
Peace in your soul for your selfish sake."
And then he stood and said was time to leave,
Said he saw nothing here that I should grieve.
Some time has now gone past,
Since Father Time and I talked at last.
The proper choices since then I've sought,
For hours and days I have thought.
Reflections of self has brought a change of heart,
I hit "refresh" and gave myself a renewed start.

I'm back in my home, so very cheery,
That heart of mine no longer sad and weary.
I have things undone, things to sew,
Old tired fingers and legs are on the go.
Once again of essence is time,
How good it feels, how very sublime.
Christmas gifts of joy to make,
Santa's back with a heavy load to take.
My presents are not wrapped in pretty paper to see,
But instead are gifts of joy and life for me.
The glow of happy faces and laughter
Will be in my heart and soul from here and after.
The sounds of love and family voices,
These are the sweet gifts of my choices.
So, Father Time, I'm sorry if I was rude and shy,
Looking back, I'm very pleased that you stopped by.
It gave me pause to see life's gifts of neglect,
It is upon those that I will forever reflect.

Sandbox

May 9, 2019

A Father's Day story

Yes, I realize it is close to Mother's Day, but this is where my mind took me today.
I don't think Mom will mind.

A boy and his sandbox toys,
A prelude to a man and life's list of joys.
Crawling for hours on knees padded with layers of patches,
Not caring if imagination and dreams would turn into matches.

Bluebird skies and a soft summer breeze
carried fragrances of lilacs and newborn leaves.
Dad was in the field cutting first crop hay;
Mom in the house cooking and cleaning still another day.

I in my sandbox had imaginations of work to get done.
Why does Daddy complain. All this farming is so much fun!
There are endless fields for my tractor and plow.
No worry if the furrows are straight, they're good enough for now.

Little did I know that these were furrows of life,
Created with passion, hard work, and some with strife.
These sandbox days were lessons well learned,
The many parts of life to come that must be earned.
Seven decades have come and gone;
My sandbox lessons are still with me at every new dawn.
I thank my dad who taught me so well to play.
I'm still in my sandbox and it's there I'll always stay.

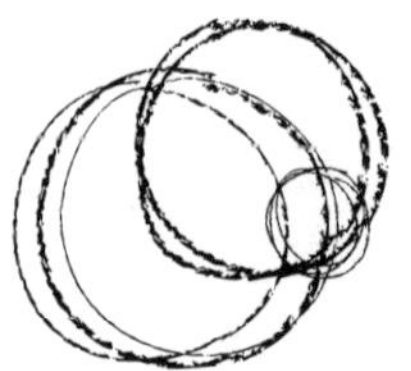

Pipe Dreams

A seemingly endless circle upon which one does trek;
Seeking, searching for a route most direct.
Around and around one goes, circling, a purpose in
life to find.
Circles of good times, others that are sad and not so kind.

What is it that one is searching for and trying to seek?
Is it fame, fortune, notoriety, or something far more meek?
There are circles for all, and all have temptations.
Which circle will grant one the most satisfying sensations?

As one travels the many circles of life,
Some are full of happiness; others full of pain and strife.
What is the pipe dream that we chase?
Maybe it is the image in the mirror that we daily face.

Upon those circles as round and round we go,
There is someone special to see, understand, and know.
Coming to peace with who and what we are,
Will give the circles new meaning, and make the trip
less far.

Silence

February 23, 2020

Silence is a mime within the depths of my solitude mind.

I hear my thoughts whispering sounds of a familiar kind.

A cool and conscious breeze hisses soft and gentle sounds.

All of life's sounds like music within my mind abounds.

There is no escaping the sounds of silence in my life.

The countless, never-ending sounds of silence dwell within me rife.

Is there truly any conscious soundless time?

Within subliminal consciousness, silence resides as a friend of mine.

I fear not the dark and emptiness of an unfounded soundless night.

The conscious thoughts of silence ever present comfort me from fright.

And with the sounds of silence in my mind I will never be alone.

Infinite silence is my soul's companion and is within me my home.

Tadpoles and Planer Boards

For David

Warm and long were the summer days.
The endless blue sky held a heavy haze.
My dad is at work and I'm home wishing,
Hoping he would come home and take me fishing.

But I decided to not sit and pout and cry,
So I went across the road to give the creek a try.
I knew nothing of a walleye lurking on a rocky shoal.
A minnow or a sunfish would fulfill my fishing goal.

I had no need for fancy fishing gear:
T-shirts, shorts, a pair of sneakers, and mud ear to ear.
At the end of a culvert was a deep and muddy pool.
I was certain it would hold something in a school.

Into the muddy pool I did wade,
A place that mom strictly forbade.
New sneakers soaked and oozing with mud.
This time I had to catch something. It couldn't be a dud.

With muddy hands and a grimy shirt,
I grabbed in murky water, with a fisherman's luck I did flirt.
And finally, a slippery, tiny creature in my hand I felt.
My heart was racing, I thought it might melt.

With this trophy treasure held tight in a slimy hand,
I raced back home to show Mom this prize so grand.
With mud from head to toe, she didn't seem to mind.
She hugged me tight, but gently and so very kind.

These days I now fish from my own boat.
But in my heart, it is the tadpole days which I gloat.
When a planer board leaps off a white-capped wave,
I wonder if it is the muddy pool that I crave?

The warm summer skies are still filled with a heavy haze,
Clouded memories to feed my wandering gaze.
These times fill my heart and soul with happy sites,
Something to ponder when I'm not getting bites.

Surf

May 18, 2020

Bold, powerful, nonconforming, symbolic of endless time;
Shaping, molding, then returning, never does it decline.
From where it comes I do not know.
Wave by wave it builds and delivers a magic show.

Listen to the songs tripping from its voice,
It is the music of the listener's choice.
At times it sings of sadness, and at other times it sings
words of cheer.
It sings in many different voices, whatever the listener
wants to hear.

It carries messages from friends and family from a
faraway shore.
It speaks of the kindness, love, and hugs we most adore.
The surf will grab your thoughts and wishes and carry
them away,
They will then fall upon another shore to brighten
up the day.

I hear the voice of the surf as it is carried on the breeze.
It speaks to me your messages as it whispers through
the trees.
I will cherish the messages the surf forever does send;
And like the surf, the messages will never end.

Temple in the Timber

When it was the springtime of my youthful years—
A time of innocence, happiness, and very few fears—
I began a journey amidst your grandeur and your charm,
Protected from the harsh winds of life, sheltered from
all harm.

Season to season we both changed and grew.
Some pieces never changing, other pieces bold and new.
I learned the hardness of the ironwood and the sweetness
of the cherry;
Saw the romance of the bee and the nectar of the
wild berry.

Decades and centuries of mysteries and stories your
voices could tell;
The truths of the memoirs no one could quell.
Nature's harps within the crown balcony play.
It is here beneath them I listen to what my God has to say.

From your hearts your sweet sap you did give;
Never complaining, providing nourishment for many to live.
Laying down your colorful canopy each season as nest;
The "Temple in the Timber," a place of comfort and
eternal rest.

The Blackboard

February 12, 2020

To my sister, with love and gratefulness.
It was just a few years ago as I recall,
Hanging left of the bathroom door on the dining room wall,
Our little blackboard hung, full of lessons to learn.
One by one, we three students each took our turn.

We started small with letters from the alphabet,
Memorizing with ease and not much sweat.
My sister, two years my senior, was my teacher.
She barked the letter to me like a gospel preacher.

I was five and she was much older...she was seven.
Little did I realize, an angel teacher sent to me
  from heaven.
She taught me the letters and how to print my name;
And to this day my signature is about the same.

The model student I most certainly was not.
My tantrums are what earned me a name called "Snot!"
But my teacher stuck with me through thick and thin.
And together, in the end, this learning game we both
did win.

Faded memories of those lessons are deep within my mind.
Seven decades later I realize my teacher was so sincere
and kind.
Through life's ups and downs, those lessons have
served me well.
They are the gift that has allowed me this story to tell.

The Test

May 7, 2020

The tests of life begin when we are born;
When from our mother's body we are torn.
It is then our hearts must first upon their own begin
  to beat.
It is from then on we recognize hunger and when to eat.

As time goes by, new challenges we must face as we grow.
We conquer them at first with instinct, just something
  we know.
But soon the challenges and the tests become to great.
We need to stretch our minds and depend more on fate.

Most of life's tests are not about being first or second best;
It is more about life after a challenge is laid to rest.
We learn of something called "faith" upon which we lean.
And after the test is when the power of faith becomes
  most keen.

The test lies within our character and our soul.
Our faith will determine the price of our toll.
So if it is a pandemic we are tested to endure,
It will be our faith that will comfort and keep us secure.

There is no need to panic or become afraid.
Just understand every test has a toll to be paid.
It is after the test we can show the wisdom earned.
Our character and faith will show what we have learned.

In a Moment

In a moment...something we often say;
A time so small, yet so big in most every way.
In a moment, we can change the way we think;
It has the power...to in a short time make a heart sink.

In a moment we can change from happy to sad,
And then, in another moment, turn back to joyful and glad.
Moment after moment is the time in which we live.
Tiny, short pieces of time put together...so much to give.

In a moment we are born and take a first breath,
And then, after decades of life...in a moment face our death.
Seemingly insignificant little moments, stitched together,
will measure all of time;
A measurement of the ages...truly sublime.

Keep It Simple

As a child, one's days are long and filled with play;
Not many worries of what others do and say.
We skip and run on lively, happy feet;
We march to a drum with a vibrant beat.

Bumps and bruises are wiped away by a soft kiss;
Tears of sadness replaced with laughter and bliss.
Life is carefree, easygoing, plain and simple.
Happy faces and grins wearing a little dimple.

Then we grow up and get a bit older;
We test our confidence...try to get bolder.
Simple times seem to get lost along the way;
Now we worry about what others think and say.

Once happy feet are often replaced with a trudge.
Happy faces are gone...now wearing a frown from a
grudge.
Where and when did the simple times get lost?
Keep it simple...losing it is not worth the cost.

The Last Hunt

It has been a while since the first hunt did start,
Guided by strong and steady hands and a giving heart.
Knowledge and tradition were topics well taught;
Nothing that was said taken for granted or for naught.

Lessons embedded firmly in a wanting, searching mind,
Waiting memories for an anxious soul to find.
Passing years and hunts building a life story to be told;
Adventures and experiences will grow and unfold.

Tracks and trails have been studied with diligence and care;
Learning to read sign...down to every little hair.
Knowing the woods and from it what to expect;
And finally, for it all...a deep and eternal respect.

The last hunt will not be the same as the first.
Hearts will not pound with anticipation...ready to burst.
Success will be measured by a wiser mind and more
weathered hand.
Stack the trophy in your memories, and let him walk away
from your stand.

When Love Is Void

So much of life passes by unnoticed...taken for granted;
Thoughts, feelings, emotions...often hidden or slanted.
When love is void, it is difficult to learn and feel.
Young hearts, soft and tender, turn into steel.

Waiting, growing, wondering of the reasons why;
Not even remembering a mother gone...not even saying
goodbye.
Now...cold and calculating is a loveless heart,
Searching, grasping for a time and place to start.

It is hard to forget the bitterness of the past.
Wondering, suppressive feelings...how long will they last?
Are the cruel, unfounded judgments just a perception?
Are the misunderstandings a misconception?

There are those who try to show relentless love...
how much they care;
Little do they know how good it would feel to with
them share.
The wall around a hardened heart where love is void,
With perseverance, precious time and love will, at last,
be destroyed.

Watch Me...Watch Me...and Keep the Yard Light On

A short story

PREFACE:

This will be an attempt to reminisce of a few days of the Christmas season as I saw it in 1955 while growing up on the family farm by the ledge near Oakfield, Wisconsin.

The days in December 1955 had grown cold, with one good measurable snowfall having arrived early in the month, and wispy flurries since to keep the wintery landscape topped with a fresh look. Life on the farm in the first winter days was usually rather mundane to a passerby, but to me it was full of recent excitement from the Thanksgiving holiday and Dad's annual pilgrimage to some faraway place in the Northwoods of Wisconsin to a place called Glidden. There, in the Chequamegon National Forest, he and my Uncle Lyle, my dad's brother, usually went deer hunting with many other relatives, that at that time, to me, were only known by name and imaginary faces as described by my father.

To say that this event of deer hunting grew as a family outing over the years to come would be a gross understatement. But, for now, it held an immediate flare for excitement and turmoil. My brother, Larry, nearly seven years my senior, was an eighth-grader, but not quite old enough to handle all the chores of milking and feeding all the cows while Dad was gone for a few days. So, Mom and Dad usually had a hired man from one of the landlord's other farms come and help.

This, to say the least, put undo stress and strain upon my mother. Mom was a bit nervous and was not accustomed to making all of the decisions around the barn, and therefore, her patience wore thin at times....well, let's say most of the time! On the Sunday night of Dad's return from hunting, it was like the second coming of Jesus to Mom, and for that matter, all of the family. We would get the chores done and then wait with bursting eagerness for Dad to drive up in front of the kitchen window, hopefully, with a big buck tied to the roof of the car. On this particular year he arrived with that big buck. My brother and I raced out through the back entryway to flood Dad with questions: "How many points? Where'd you hit him? Did you have to track him far?" and on and on and on! Even though I was only six years old, I had the hunting jargon down to a tee. My brother and sister, Judy, and myself, had heard the stories of all Dad's hunting trips a thousand times, I am sure. To this day I can tell them all just as if I had been there myself. To do so is sacred to a deer hunter. There is an unwritten fraternal bond in the deer hunter's world. Stories are repeated to instill accuracy in the listeners to preserve tradition. Little did we know that 50–60 years later we would have that story and many, many more memorized from years of reliving them as we told them over and over.

At the tender age of six, I was still a "wannabe" believer in Santa Claus, and thus, it would only take minimal care and effort to hold me to the "believer" side of the line! This is where the deer entered the picture as a significant and integral part of Christmas. Dad would save the legs from the deer after butchering and use them later to make tracks in the snow from Santa's sleigh and reindeer when he came on

Christmas Eve. I don't remember at what age I quit buying into it, but to this day I'll confess to a tinge of excitement when I see deer tracks in the snow!

School days were, likewise, filled with activities and preparation for the holidays. It should be noted that this is the only year that my brother and I attended school together, and, sadly, I don't remember much about Larry being there. Each year our teacher would prepare a Christmas program for children in first grade through eighth grade to present to our parents. There were usually 20–25 students in those eight grades in a one-room school with a wood or oil furnace and outhouses for bathrooms! The program involved short, memorized speeches, songs, and short skits of the seasonal theme. My lines were often long and arduous, the likes of which would turn any ordinary brain to mush! Like, "I had a little calf, and that's half; I put him in the stall, and that's all." The thought of having to stand in front of family and neighbors and try to recite sends chills up my spine to this and my very dying day! Can you imagine this scene today... one teacher, teaching all eight grades, no plumbing, firing a wood stove, and preparing a program in a public school, pledging allegiance to our national flag, and disciplining with a firm hand...or a yard stick! Heaven forebid, it's a miracle we all survived. We did, and some became doctors, lawyers, businessmen and women.

Dad was clerk of the school district at Genesee Public School, of which I have been writing. Our Christmas program would usually conclude with the parents holding an auction. It was called a "box social" in which each of the mothers would pack a box lunch to be auctioned off to the husbands. Which lunch a husband purchased he had to share with the mom who had made the lunch. My dad was always the auctioneer, and it was always interesting to see the partners. I'm sure what I didn't know was that there were probably a great many adult-type innuendos in the matchups! The passing of this program meant it would only be a day or two until Christmas Eve and Santa.

Christmas Eve day was always special...filled with family laughter and wide eyes anticipating the coming of Santa. I was always careful to make certain to do all my chores with haste and willingness, just in case Santa had a helper watching...which he did. He was wearing Oshkosh B'gosh bib overalls, and he was milking cows. What a conspirator! Those evening chores were nearly impossible to get done. The excitement of the moment would have my heart pounding like war drums. Fortunately, I don't remember why, when, or how, but Dad would sneak away from milking for a few minutes, grab the deer feet and something to make sleigh marks in the snow, and make the pure evidence of Santa's arrival for any and all to see. One could not doubt or contest the evidence. Those tracks would be on the "airing porch" upstairs where my grandma lived, and they would lead to the roof in front of Grandma's kitchen window where the cats always sunned themselves. It had to be real. It was there. I couldn't see way up on the roof, but Dad said they headed straight for the chimney, and that was good enough for me. I spun around like a top, too fast for undersized feet and oversized boots of the hand-me-down nature to stay headed in the same direction. Someplace between the southeast corner of the porch and the top of the stairs my boots caught up to my feet, and in my mind's eye I descended those long stairs with the speed of Hailey's Comet! I grabbed frantically for the handle on the old wood screen door, and then for the brown, glassy knob on the inside door. Once inside, with Dad and Larry and Judy close behind, I twisted my bony body inside a much too large jacket that should have been at waist length but, on me, hung halfway between my knees and ankles. On the farm, things were of purpose, not of style. On the north wall was a long one-by-six board nailed to the wall, about eight feet long, with a row of the screw-in hooks to hold our coats and caps. Beneath that, about three feet up from the floor was another board, put there, I'm sure, for the smaller people to use. With a little help from Dad, I had shed my wintery wardrobe and was headed for the bathroom, where I'm certain Mom was going to intercept me, give me a quick wash job, and PJs.

My grandma lived upstairs in our farmhouse. That part of the house was her home, which originally had been for her and my grandpa Alfred, who died before I was born. I have always felt bad, to this day, that I never knew him in person, only in family stories and photographs. He has always been described as powerfully built and of the highest character. He immigrated to America from England when he was a very young child. I carry his name as my middle name and do so with great pride.

Having torn from Mom's grasp, one ear short of being clean, she told me to tell Grandma to come downstairs to open presents. The pajamas I wore were of the two-piece variety, medium blue as I recall, with orangeish-colored tigers on them. But, like the rest of my clothes, they were second or third hand, and in my case a little oversized. Mom always said that was good because we could grow into them, which I finally did in about sixth grade! I rounded the corner between the old kitchen table and the stove and circumvented the stairs in the most awkward style...trying to run and trying to hold the pant legs from tripping me and at the same time trying to keep my bony skeletal behind from over-exposure. Grandma was sitting in her usual spot in the corner to the left as I entered her living room. She held in her lap the wicker style oval-shaped basket that contained a couple spools of threadthread—one white, the other black. In her hands were a darning needle and a wooden darning spool with a sock stretched over it. This was standard procedure in those days as clothes, including socks, were mended time after time until there was no room for more patches. Grandma didn't need any further instruction, as my mere presence was convincing of the business at hand.

Downstairs, our living room held the tree, decorated... mostly by mom. Dad usually brought the tree home from up north when he returned from deer hunting. Mom was a perfectionist, and the tree always displayed her talents with tinsel and every ornament placed precisely. Under the tree was a bounty of gifts, waiting to be claimed by an anxious, but grateful recipient. Mom sat on the floor, legs crossed in her usual manner, digging through the presents, looking for one for Grandma because we always started with the oldest first to unwrap. I don't recall what Grandma, Dad or Mom got, but I do know that most gifts were items of need, not necessarily want. Being the youngest, I realize now, that oftentimes the rule was overlooked in the case of my presents. I think this was the year that my brother built a new barn to use in my sand box. It was tremendous! It was designed and put together to look almost like our real barn on the farm. It had a round type tin roof similar to the round rafters of the barn Dad would have in the future. The walls were white and made of wood. I had that barn for years, until I outgrew playing in the sandbox, and then it continued to employ the imagination of my nephews years later. Gifts from Santa mostly were for the youngest members of the family. My brother and sister would usually get something of need, like socks or underwear, gloves, mittens, or a shirt. One year, it might have been this year, Larry got a new .22 caliber. Remington rifle, which at age seventy-nine, he still has in perfect condition. It never seemed to make a difference as to what we got if one gift was more than another. We were just happy to get something. With the gifts all open, the rest of the evening was spent trying on new clothes, playing with a new game or toy, and eating some special hand-packed chocolate candy from our landlord. Finally, it was off to bed to try to rest for the following day, Christmas Day.

Christmas Day was spent with relatives and consisted of a major dinner, including ham or turkey, sometimes both. The responsibility of it all was moved from one family to another year after year. I cannot remember where it was in 1955. The day consisted of the dinner, a gift exchange based on name exchanges previously done at Thanksgiving. The gifts were opened after the dinner, and the gift items were often not things of much expense or value or need. Being only six, I did not grasp the concept of the exchange, nor did I care! It was just something fun, and it didn't need justification or reason. Oftentimes, the gifts my mom or dad or aunts and uncles would exchange brought about tons of laughter. Later in life, I would learn why, as many of the items exchanged were pranks of silliness. The important thing was it kept all our families together, something somewhat missing in the modern era. One gift that I remember was a rolling pin. I am certain that by this year, 2021, every aunt, uncle, cousin, niece, nephew, and probably their children, have had the rolling pin. It just keeps rolling around! Somehow, I'll probably regret mentioning this.

Our day of celebration ended in late afternoon when we all went our separate ways to do farm evening chores. During this era we were all farm folks, something that would change in generations to come. My sister and brother both had to help with all tasks at hand in the barn, as did I. Size or age did not relieve one of responsibilities. Upon arriving home, good dress clothes were exchanged for old familiar and smelly barn clothes. It was the only time I would realize the smell of the barn and cow poop. Otherwise, one became immune to those smells. To me, the barn stood like a monument. It was a symbol of strength and security...like a great Egyptian pyramid.

It felt comforting for some unknown reason to be back home and get back to our chores. My brother would go about throwing down silage from the silo, while my sister and I would fill the feed cart and start feeding two rows of cows...27 to a row. The cows faced outward in our barn and, therefore, had a view out the windows. I have never thought about that until now. It should have been better than facing in toward each other, not a face I'd want to stare at all day and night! Judy would take the cart and push and I would get on the opposite end to pull. Sometimes it would lead to heated arguments as to who was helping or hindering the operation. When we were out of control, Dad would intervene and we would have to sit on a pail or on the cement curb in the ally leading to the milk house. On this day, for some reason, I was sent to the west end of the barn for my punishment and told to sit on a bale of straw in front of the lime bin. The lime bin was along the north wall in the lower, modern addition to the original barn. It was where the harnesses were hung when Dad still had draft horses. I recall kneeling on patched knees and playing with two kittens. I was folding pieces of straw into little squares like imaginary packages for the kittens to open. Razor sharp claws would get caught in my lime-green wool mittens...mittens made by my grandma Johnson, my mom's mom. As I played and retrieved my mittens from the furry pranksters, I could feel my toe pushing through the mended sock I wore. I guess Dad could sense the fact that I was growing tired from the long day and suggested I go to the house and get ready for bed. Under most circumstances, I probably would have fought the idea, but on this day it seemed fine with me.

I began the walk to the milk house to make sure I would meet Larry along the way. I needed him to help get me to the house. I could not go to the house in the dark, alone, without being watched. There were boogeymen, and that was a fact. I had heard my mother refer to them on several occasions when I was naughty. She would not lie. They were real. They were there. She even knew the noises they made. Sometimes we could sneak up behind her and scare her by hollering Boo, and it would scare the bejesus out of her. I've seen her nearly jump out of her skin. In later years she gained weight and said it was all because of being scared too often, and nearly jumping out of her skin, stretching it so that she had to eat more to take out the wrinkles. Now I know what she meant because I'm getting some of those wrinkles too, and I'm gaining a little weight trying to keep up with them. At any rate, my sister could not be trusted. She would, without fail, shut off the yard light, thus allowing boogeymen to come from out of nowhere!

In 1955, the closest thing to Air Jordans was PF Flyers, of which I did not own. Remember, I had on those hand-me-down boots, three sizes too big, the long jacket below my knees, and a cap that pushed down my ears like miniature wings. I was not a picture of speed. I was no match for the boogeymen, or so they thought. And thus, I would plead, "Watch me. Watch me and leave the yard light on." Then, with all the power and will a six-year-old, somewhat anorexic, scarecrow bag of bones could muster, I would leave the milk house steps and begin to run to the house. It was a sight for sore eyes to see...with the oversized boots still standing on the steps, my feet and body...thirty feet away, past the silo before the boots and jacket caught up. Stopping was not an option. I was going through the back entryway door of the house...open or not. Somehow, I always made it...opening both entryway doors, in a move to this day unknown to any other human.

This year, 2021, the barn that was a monument in my eyes has been long demolished. It will always stand in my mind. It has been sad to see it crumble under the jaws of backhoes and bulldozers, helpless to the progression of time. This Christmas Eve, for something different, but familiar, maybe we can all go to the barn again, and when chores are near done, and it's time for me to go, and I could say, "Watch me. Watch me and leave the yard light on."

When Spirits Meet

November 2019

This is a story of my hunting of the great white-tailed deer.
I believe his Spirit will hear me tell the story.

A final time I came to visit among your enchanted trees;
As you can see I am stationed up high near the honey bees.
This year I announced my entrance with my usual noisy
clatter;
To be stealthy and quiet, no longer does it matter.

I have been chasing you for decades in many
different places.
So difficult to find and see, your spirit wears so many faces.
You used to hide in the field of corn on my father's farm,
But you were wise and used a vanishing guise to sly
away from harm.

This time my search is of a much different quest,
I have been invited to your woods as an honorary guest.
No longer is it my hungry ego that I wish to fulfill;
I want to understand and capture your spirit and your will.

And God willing when at last our spirits meet,
We will both at last have conquered our final glorious feat.
Soon our spirits will be in total harmony high up in our sky,
And no longer will we question each other of the
reason why.

When We Are Called

November 2019

There are times in life when a "calling" for help feels
to be unfair.
It fills our lives with turmoil and uncharted and
unwanted despair.
We have reached our golden years so it is they say,
But when we searched our soul's bank, it was not
prepared to pay.
We searched our pockets for strength, advice, and will,
But found for this burden and pain there was no magic pill.
There is but one place left for help and that lies within
our heart.
So it is there we have found the help to do our
needed part.
We have come to realize there needs to be no reason why.
We were called because of needed help, and we are
here to try.
Each day we awake and try our best to do that which
is just and right,
Making sure it is our love and faith that we keep within
our sight.
So on this day of peace and joy and grateful thanksgiving,
It is the burden of the calling we are most forgiving.
We no longer feel the pain or the extra work and strife.
We only share the thankfulness that we were Called
to help save a life.

Willow

Willow, Willow, always bending for me.
It is not your weeping that I see.
No, no, your long and slender leaf,
Has only shades of peace and comfort, none of grief.

Willow, Willow, with gentle nodding crowns,
You beam with only joy and happiness and whisk away
  our frowns.
On your delicate branches our memories ride,
And beneath them, all our wishes we confide.

Willow, Willow, weep not for me.
Winds on my cheeks kiss me softly and tenderly.
Though your limbs may droop and fall,
Willow, Willow, you are my oak, standing proud and tall.

Winter Wind

December 20, 2019

Emboldened, brazened, sharpened is your angry scowl.
Beneath your seeming wretchedness follows your embittered howl.
Is there nothing but pain you carry on your blowing face?
Or is there a softer, hidden truth of a more gentle, shameless grace?
Where lies the secret messages of your frozen voice?
Does your singing truth fall upon my soul's wanting choice?
There lurks a silent softness deep within your frightful, icy face.
Maybe it is my vulnerable imagination that I let you chase.
Are the answers in your naked drifts of pristine snow?
The messages may speak from the crystal sparkles that upon them glow.
No longer will it be your frigid scowl or your shrieking howl that I most fear.
It is the purring, peaceful grace of the winter wind that I will hear.

Acknowledgments

I want to thank Tanya, my wife, for her endless, timeless support, for the book creation, and my everyday life. Her moral and physical efforts were an integral part in creating this book. She was my first read and provided all the technical support organizing the written documents into presentable material to present to the publisher. Without her knowledge and dedicated support, this book would have never happened.

Special gratitude to my brother, Larry K. Binning, Professor Emeritus, and his wife, Dorothy, for reading the documents, one by one as I wrote them. Also, for their support and encouragement, and for getting the idea of a book off the ground by presenting the documents to his friend and colleague, author Jerry Apps, Professor Emeritus, for review and comments.

Thanks to my sister, Judith Renderman, and her late husband, Stan. Without her good and fond memories of our family's lives, I would have been lost searching for details of events of our past. Also, memories of working with her husband, Stan, and years of shared hunting trips together, were instrumental in creation of the book.

To my sons, David Binning and Jeff Binning, I wish to thank them for all the fond memories as we all grew together. These memories are a great part of the platform from which I have written this book.

Lastly, but certainly with deep respect, love, and indebtedness, thanks to Robin and Brian Willett, my wife's parents.

Without their generosity and love and kindness, I would not have had the time and initiative to dedicate to writing this book. Also, to my sisters-in-law Pam Willett and Maria and Maria's husband, Danny Vrana, thank you for your never-wavering kindness and inspiration.

To all the folks not mentioned by name—friends and neighbors—you know who you are and how much you mean to me and my life. Thanks for always being there for me.

About the Author

Wayne A. Binning is a native of the Oakfield, Wisconsin, area, where he was raised on a small dairy farm and attended Genesee Public School for eight years. He graduated from Oakfield High School in 1967. He gained his formal education at what is now the University of Wisconsin-Stevens Point. Today, he is retired after spending most of his career as a field supervisor in the canned vegetable industry. He currently lives in north central Wisconsin between Wausau and Antigo. He and his wife, Tanya, enjoy the wooded peaceful setting in which they live and watch all of nature's critters just outside their door.

Printed in the United States
by Baker & Taylor Publisher Services